SQUARE IS A SHAPE

a book about shapes

by Sharon Lerner

Lerner Publications Company, Minneapolis, Minnesota

To Rachel

Third Printing 1974
Second Printing 1971
Copyright © 1970 by Lerner Publications Company

All rights reserved. International copyright secured. Manufactured in the United States of America. Published simultaneously in Canada by J. M. Dent & Sons Ltd., Don Mills, Ontario.

International Standard Book Number: 0-8225-0272-0
Library of Congress Catalog Card Number: 70-91671

These are shapes.
There are many different shapes.

This is a circle.

This is a square.

This is a triangle.

Some triangles are tall and thin.
Others are short and fat.

Two squares make a rectangle.

Two triangles make a diamond.

Three squares make a longer rectangle.

In the circle family there is

the oval,   the crescent, the arc,

and the semi-circle.

Shapes are like blocks.
From a pile of blocks you can build many things.

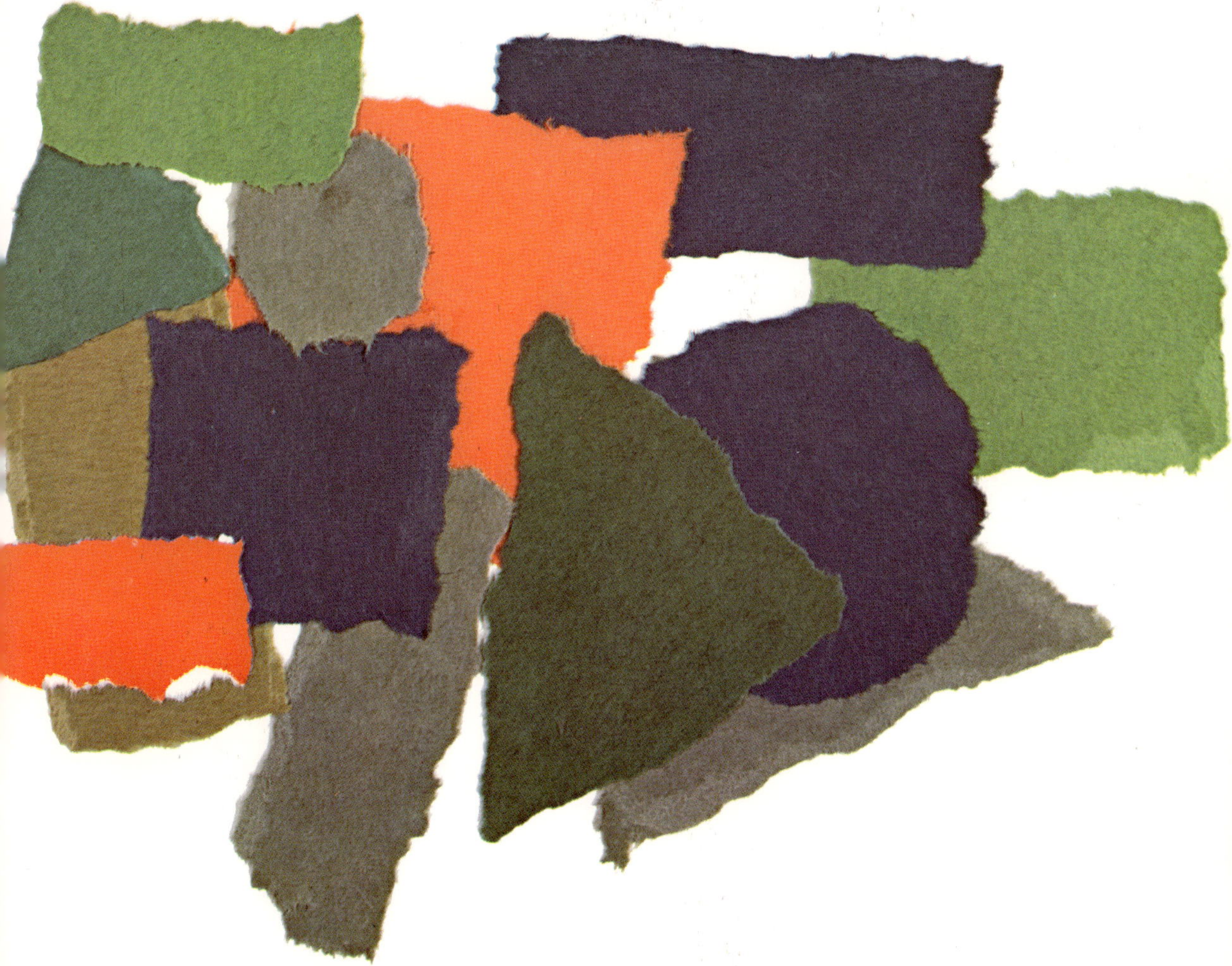

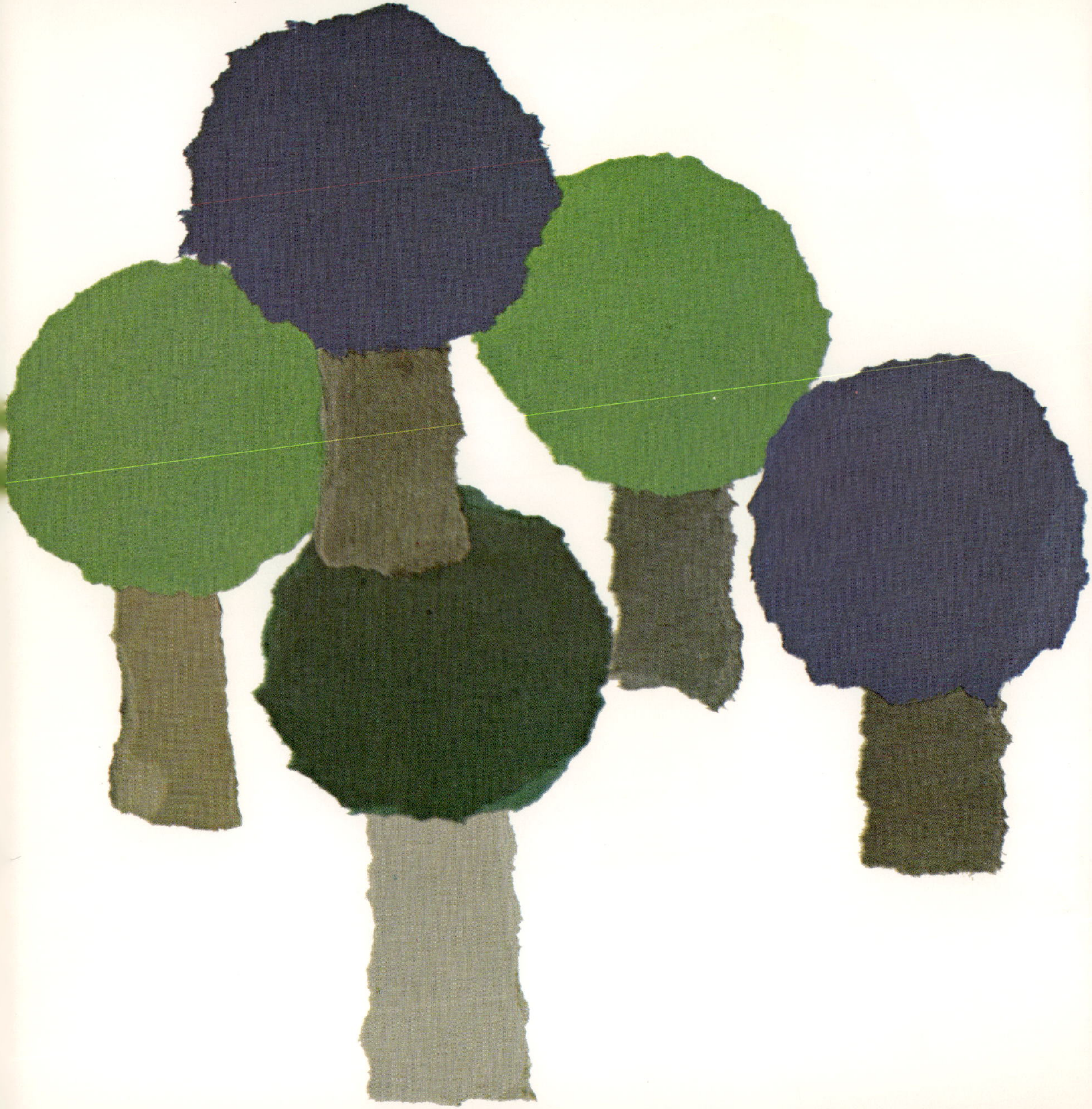

Circles and rectangles become trees in the woods.

Semi-circles on triangles turn into ice cream cones.

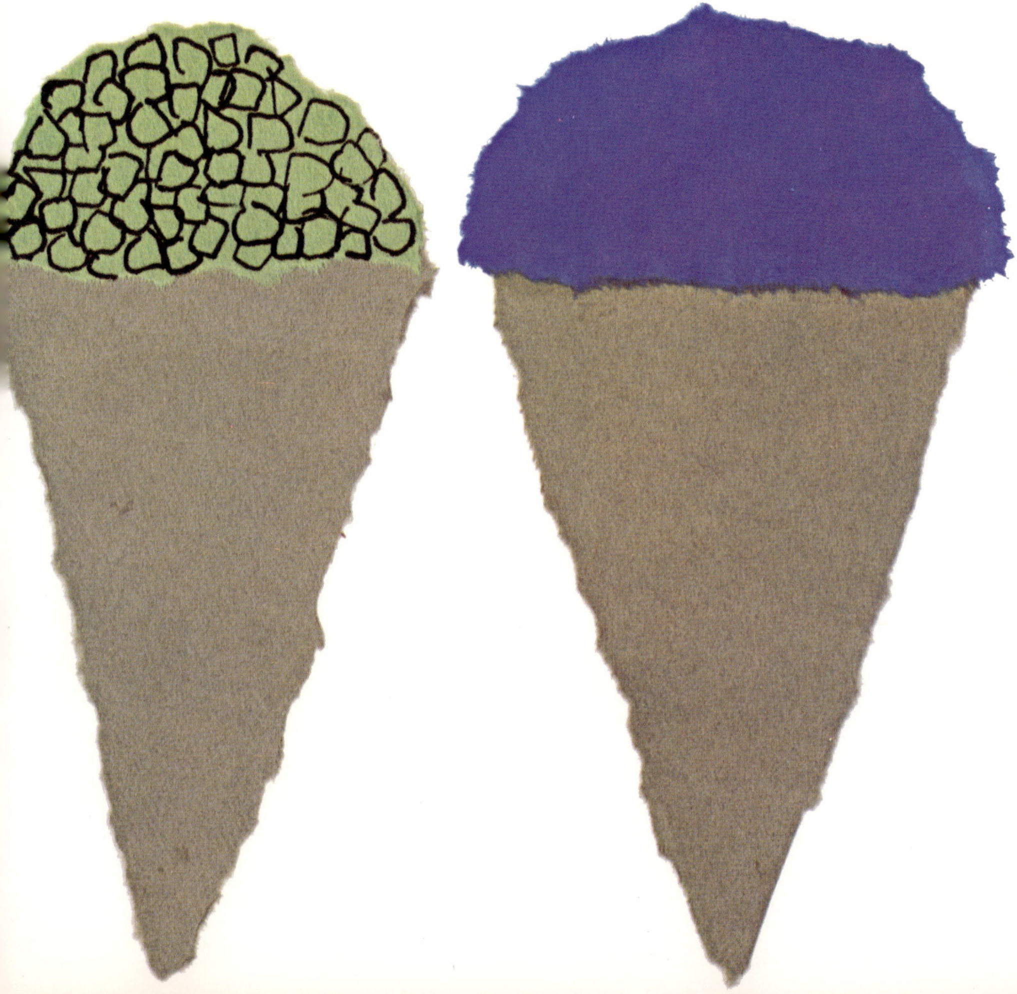

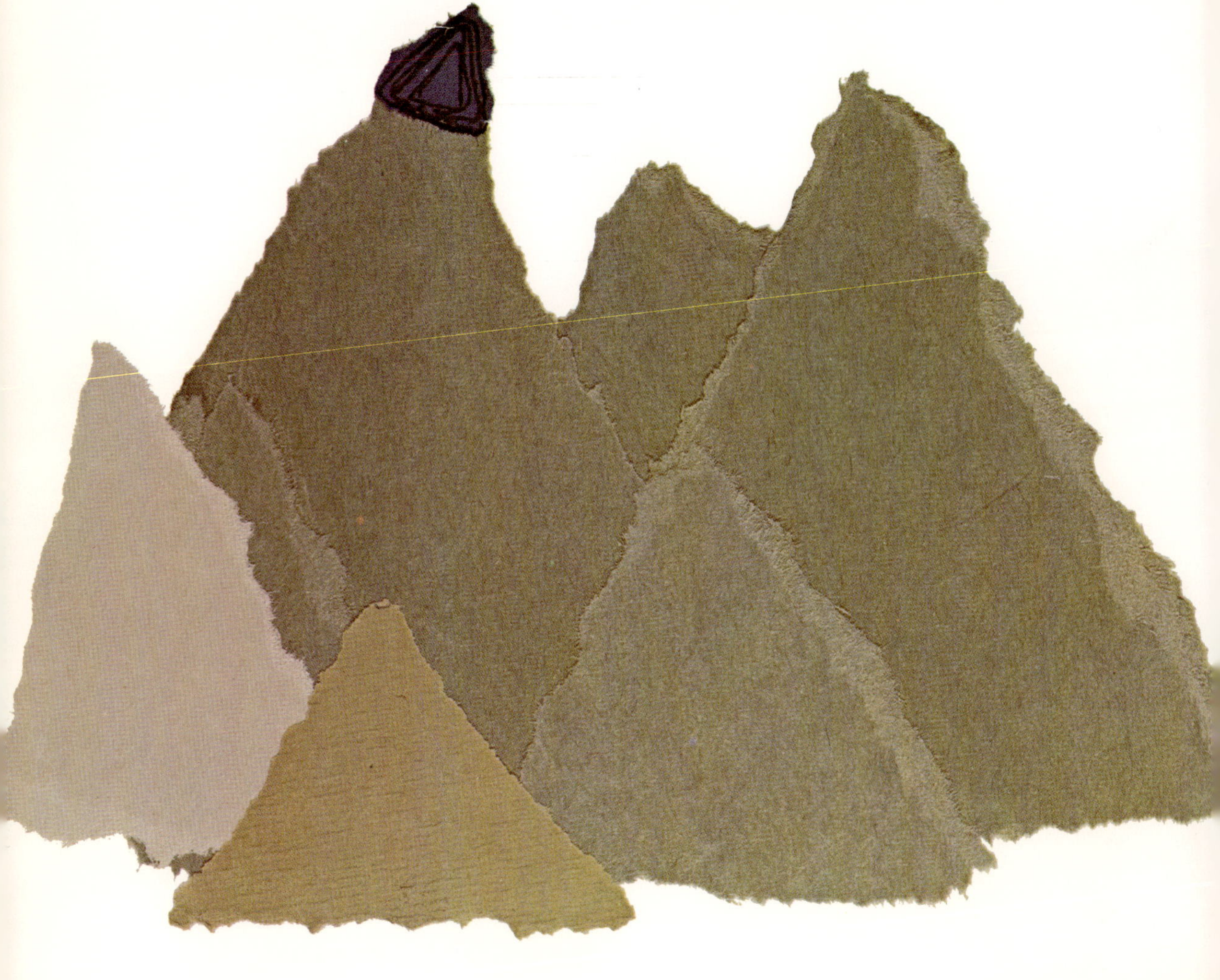

A line of triangles is a chain of mountains.

A square is a page of this book.

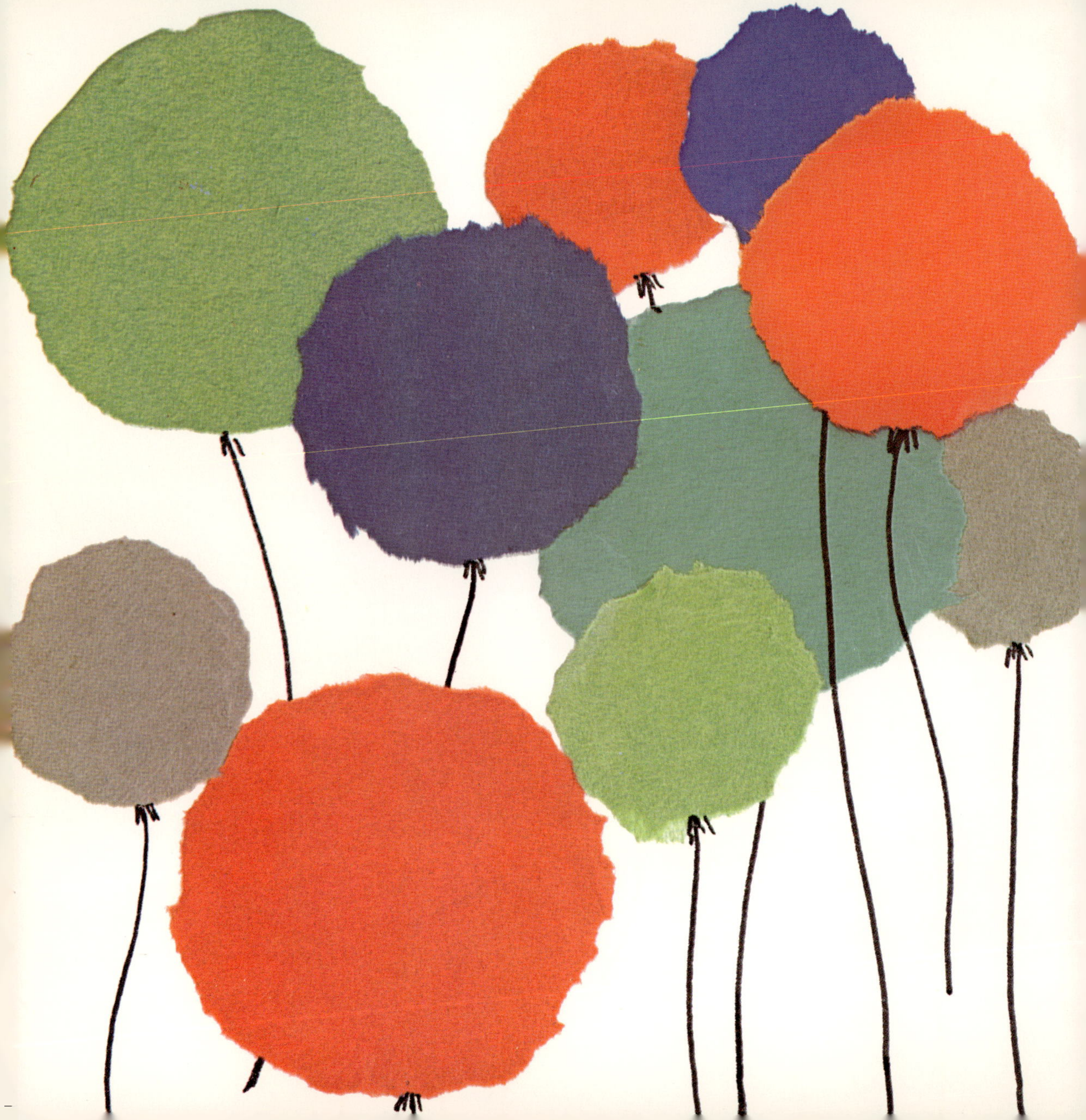

Circles on strings are a bunch of balloons.

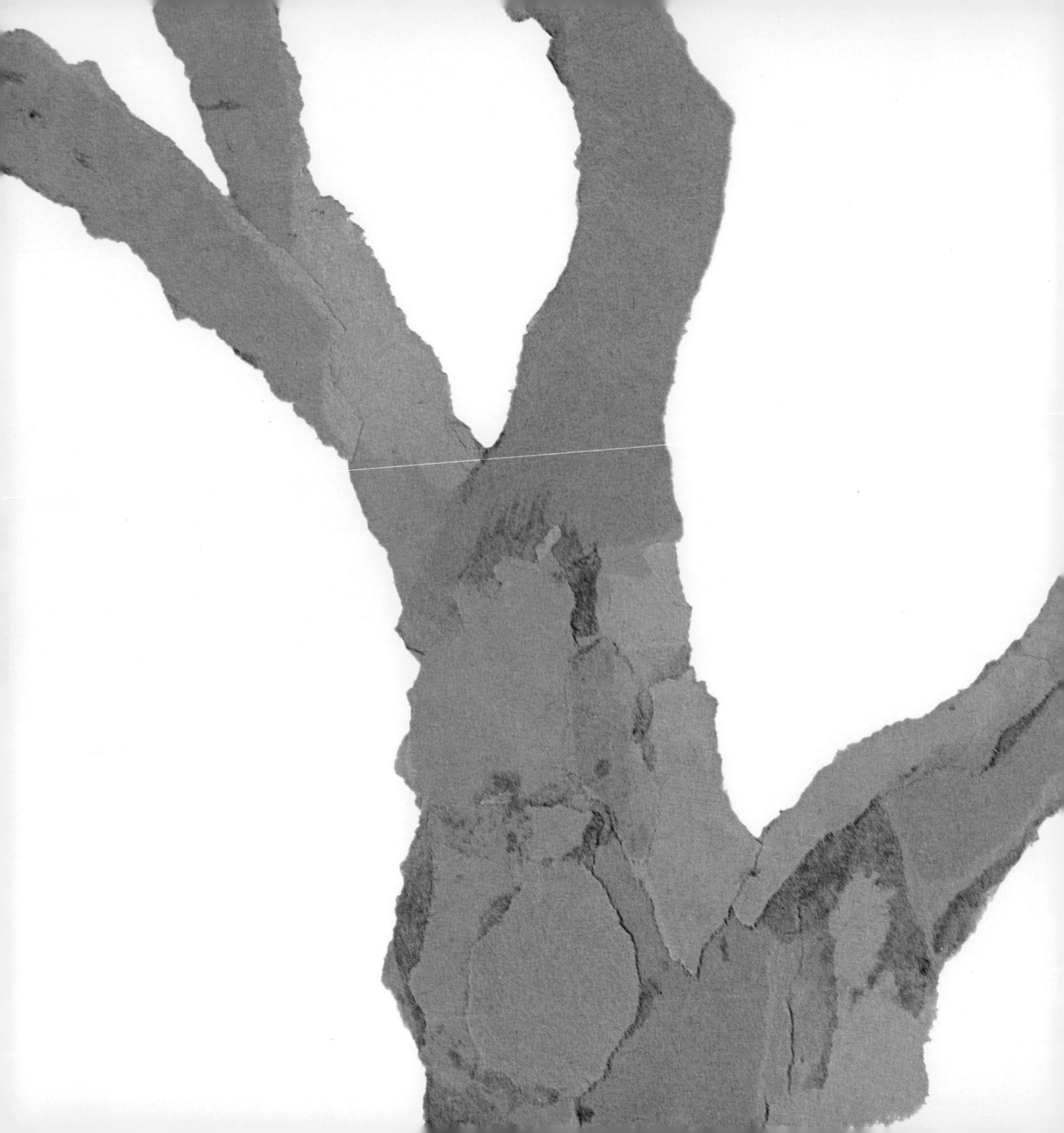

A square and a triangle make a brown wren's house.
A circle is his door.

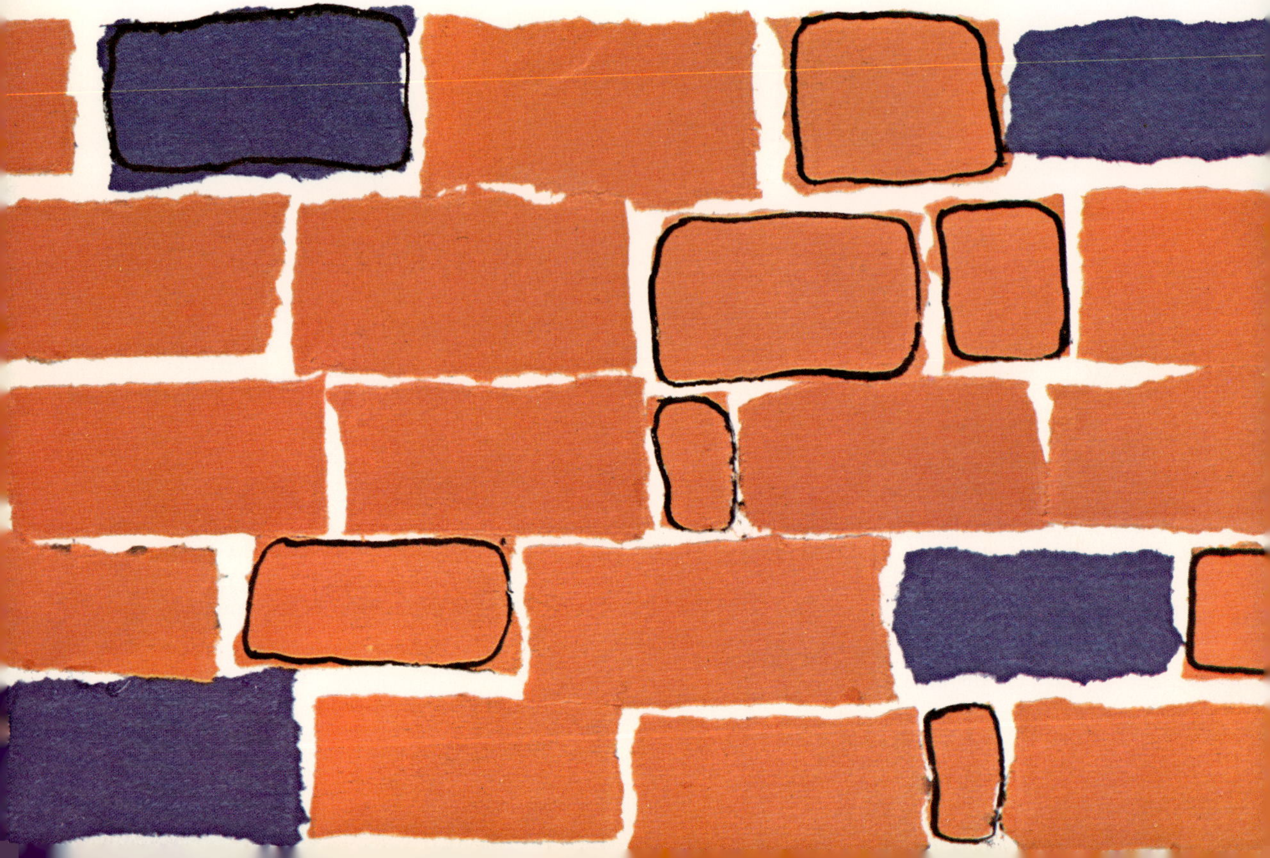

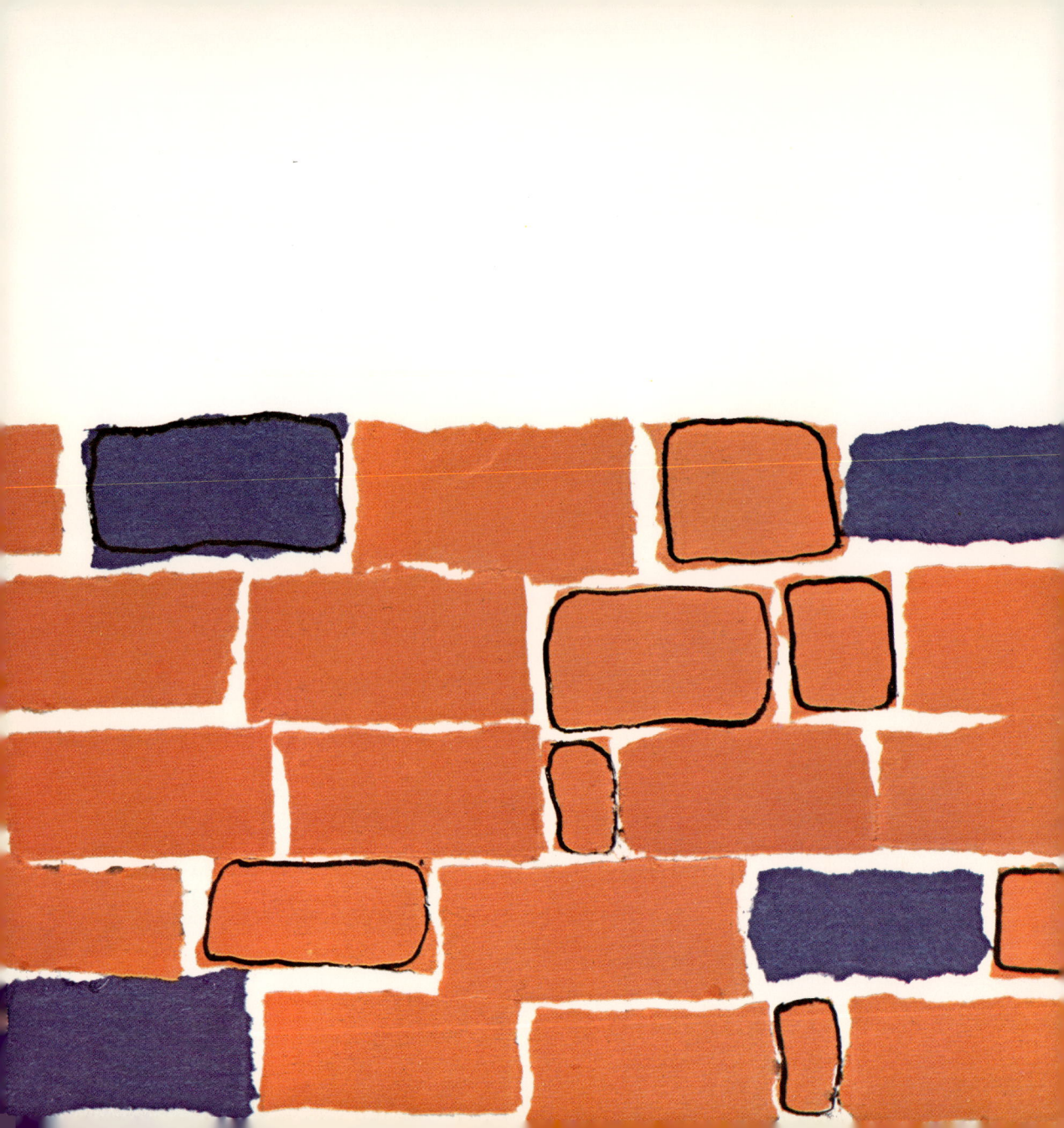

Row upon row of rectangles and squares make a good wall.

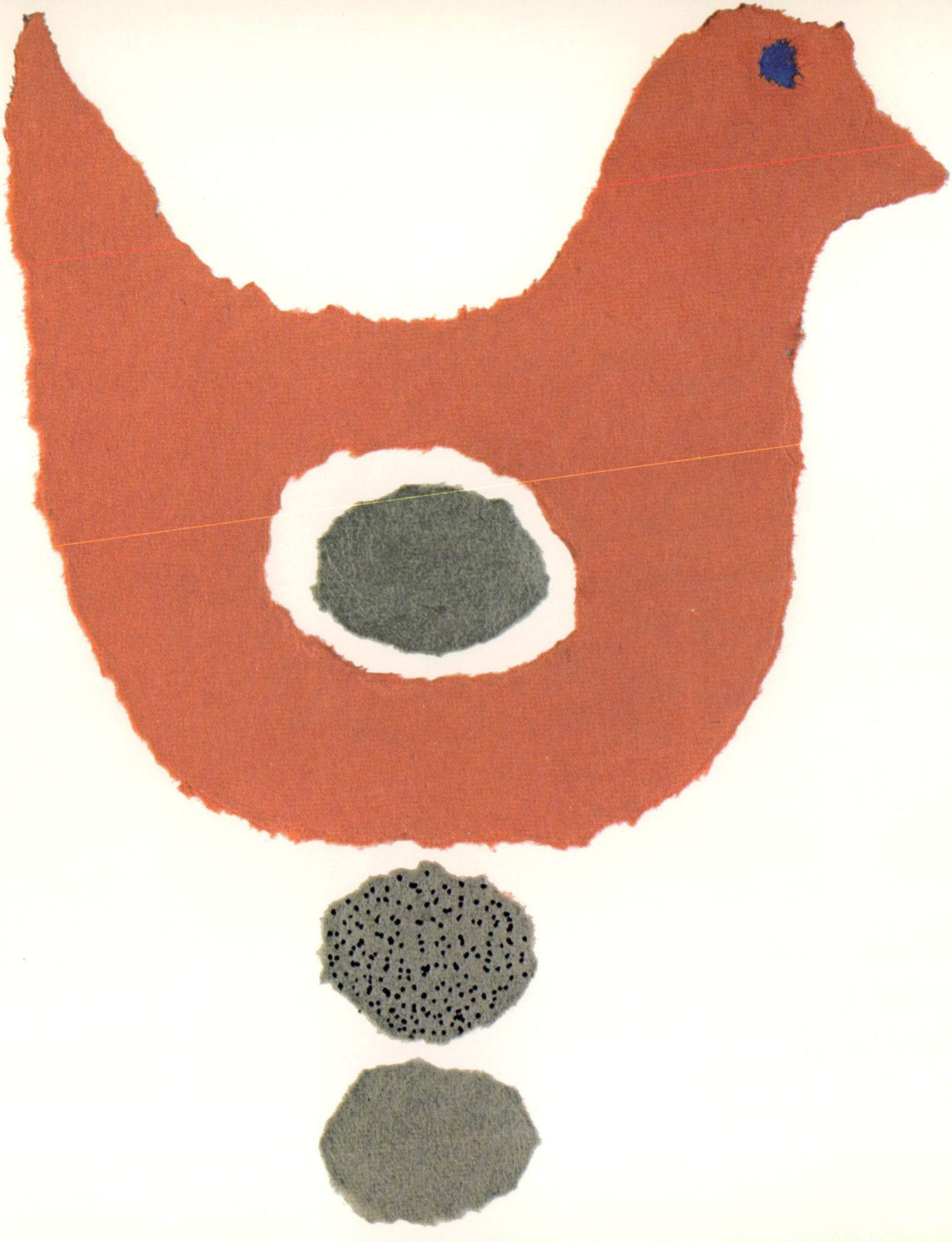

A chicken lays ovals.

Even a train is made out of shapes

from beginning to end.

ABOUT THE AUTHOR

Sharon Lerner's published works combine her love of nature, art, and writing. As an artist, Mrs. Lerner has been recognized for her watercolors, collages and jewelry. She has a degree in art education from the University of Minnesota and has been a lecturer and guide at the Walker Art Center and the Minneapolis Institute of Arts. She has taught at University High School, Walker Art Center, and the White Bear Public School System. Mrs. Lerner is an experienced writer and illustrator whose books include *Places of Musical Fame, The Self-Portrait in Art, I Found a Leaf, I Like Fruit, I Like Vegetables, I Picked a Flower, Who Will Wake Up Spring?, Square is a Shape, Orange is a Color, Straight is a Line* and *Butterflies are Beautiful.* She lives in Minneapolis with her husband and three children.